Tough and scaly, reptiles roam the Ea[...]
longer rule it as the dinosaurs once di[...]
live in many different habitats, from ocean to river,
from desert to tropical rain forest. They were the
first *vertebrates* (animals with backbones) to lay eggs
covered by a shell, allowing them to survive on land.
Being "cold-blooded," modern reptiles adjust their
body temperature by seeking places either to warm
up or cool off.

One major group of reptiles includes the
tuatara (too-uh-TAR-uh) from the islands
off the coast of New Zealand. Here, a
tuatara comes out of its burrow to look
for food. Tuataras have changed little in
over 200 million years. Slow to grow,
tuataras live about as long as people do.

Crocodiles ~ Gharials ~ Alligators ~ Caimans

It's hard to tell where the streamlined body of a crocodilian (krah-kuh-DI-lee-un) ends and its powerful tail begins. In water the crocodilian moves by swinging its tail, paddle-like, from side to side. All crocodilians are predators. Their long snouts, strong jaws, and pointed teeth make it easy for them to grab hold of their prey and rip it apart. Like all reptiles, crocodilians need to warm up and cool down, too.

The Siamese crocodile is one of about 25 types of crocodilians found mainly in warm, tropical climates. Here, a group of them gathers near the water's edge.

A plump gharial (GER-ee-el) holds its slender mouth open to let out body heat and cool down. It hunts for fish in the rivers of India.

Although this American alligator is basking in the sun, it can survive colder temperatures than crocodiles and gharials. Can you see the bony armor on its back? It protects crocodilians from harm.

This young spectacled caiman (KAY-mun) floats on the surface of a flooded Amazon forest, waiting for a chance encounter with a nighttime snack.

CHELONIANS

Tortoises ~ Turtles

Tortoises take to land, but turtles favor water. Sandwiched for life, chelonians (kih-LO-nee-uns) carry their protective housing wherever they go. The shell is fused to vertebrae and ribs, and grows with the rest of the body. The largest adult turtle has a shell more than 6 feet long, and the smallest has one less than 3 inches. Although they have no teeth, their strong jaws grasp and slice food easily.

Galápagos (guh-LAH-puh-gus) tortoises lumber through their volcanic island habitat and take a dip in a pond. Those on some islands have very long necks that can reach up to feed on tall cactus plants.

Marine turtles—like this beautiful green sea turtle—spend their entire lives in the ocean, only leaving water long enough to lay eggs in the sand.

KEEP AWAY! A red-eared slider peaks out from under its shell. As it hisses a loud warning, its horny beak reveals a toothless mouth. Fossils show that some chelonians had teeth; chelonians today do not and must snap off pieces of food and swallow them whole.

The alligator snapping turtle can bite with crushing force. When it submerges, the pink worm-like lure on its tongue wiggles to attract fish. But it won't be a fish who gets to *eat* dinner!

SNAKES

One type of reptile—the snake—has no legs, eyelids, or outer ears, but makes up for what it doesn't have in other ways. Its long, slender shape helps it slip into tight spaces. Some snakes even have special teeth, called fangs, to inject venom into their prey.

A long, colorful snake from Panama glides up a cluster of plants. What it lacks in keen eyesight, it makes up for with its forked tongue, which is sensitive to odors and taste.

When the Cape cobra is frightened or ready to strike, it flares its impressive hood.

At home in the Australian Great Barrier Reef, this sea snake seems to move effortlessly through the water, aided by a long, flattened body and paddle-like tail.

Like all snakes, the smooth green snake must shed its skin as it grows. While shedding, its papery old skin turns inside out. Even the outer layer of its lidless eyes peels off.

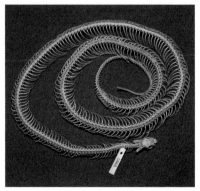

Many vertebrae and ribs give a snake great flexibility to move in any direction. A human backbone has 32 vertebrae compared to as many as 435 in some snakes!

A biologist "milks" venom from a rattlesnake's fangs. It will be used to make anti-venom to help snake-bite victims. But when the snake injects venom into prey, it destroys the prey's muscle and other tissues, preventing escape.

LIZARDS

From the smallest to the largest, lizards range in size from the 1 1/3-inch Monito gecko to the 10-foot Komodo dragon. They come in many forms and colors, but share a few features in common. They have legs and ears, and most have eyelids. More than 3,700 species live in varied habitats—from the tropics to much colder climates, from sea level to the mountains.

The Texas horned lizard inhabits hot, dry places. It lifts its body well above the scorching sand to stay cool. Prickly skin discourages would-be predators.

One of two poisonous species of lizards in the world, a Mexican beaded lizard emerges from a woody nook. When this lizard bites its prey, venom oozes from glands above the teeth and enters the wound, paralyzing the victim.

ZAP! A Parson's chameleon (kuh-MEEL-yuhn) shoots out a sticky tongue—longer than its own body—to catch a cricket. Its eyes can move independently, but they also work together to focus on prey.

While this reptile might look like a snake, it's actually a lizard. The glass lizard has ear holes, eyelids, and tiny hind legs. Its smooth-as-glass surface helps it burrow in the ground.

This Galápagos land iguana rolls the spines off a prickly pear fruit before eating it.

Reptiles move in many ways. Whether they burrow, glide, slither, or swim, reptiles are well suited to get around!

A frightened basilisk lizard dashes across the water's surface with the help of long limbs, flattened feet, and its tail for balance.

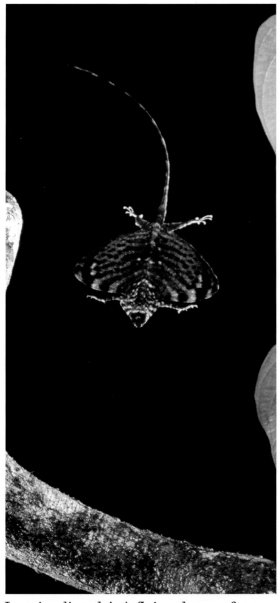

Powerful flippers propel marine turtles—like this green sea turtle—through the water.

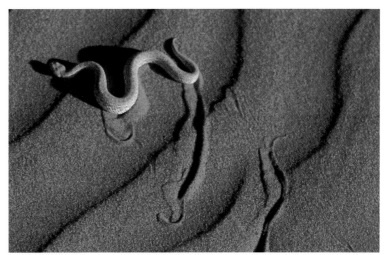

Leaping lizards! A flying dragon from Indonesia soars between treetops. How? It unfolds its long, hinged ribs that open like wings to glide through the air.

"Snaking" across a sand dune, the sidewinding adder throws its body forward in long loops as its tail pushes against the sand from behind. Sidewinding is just one way snakes get around.

A Tokay gecko walks upside down along a branch. By curling back its toes, the gecko lets go and takes another step.

A close-up look at one of the gecko's feet reveals how it hangs on. Rows of microscopic "hairs" on its wide toes stick to the surface.

Why do certain animals put on a great show with wild colors or bizarre disguises? Sometimes it's to compete, sometimes to attract a mate, or maybe to scare off something bigger than they are.

Two male anole (uh-NOH-lee) lizards signal with their flashy throat patches, called *dewlaps*. Males compete with each other for territory and females. A female will pay close attention to the display.

Throwing its head back and erecting a huge membranous collar, the frilled lizard appears more ferocious and larger than it actually is. It behaves this way when frightened.

With its three bony horns, a male Jackson's chameleon resembles a rhinoceros. Males fight and can lock horns as they battle over a female.

REPRODUCTION

Unlike amphibians, which deposit jelly-like eggs in water, reptiles lay their eggs on land. They can do this because their eggs are encased in a shell, which keeps the *embryo,* or developing egg, moist. A number of turtles, lizards, and crocodilians develop into males or females depending on the temperature at which the egg stays. Some reptilian parents guard their eggs and young, while others do not.

A Nile crocodile emerges from its leathery egg.

A few reptiles, like this eyelash viper, give live birth. Do you see the golden and mottled babies?

A green sea turtle lays eggs in the sand. Then she covers them and returns to the water, leaving the young to fend for themselves.

A green python wraps her body around her eggs. By shivering, she raises the temperature of her muscles, and the heat warms up the brood.

Protective parents, crocodilians guard their young and respond to their calls.

HUNTING AND DEFENSE

To eat but not be eaten can be risky business for reptiles. Some hunt by sneaking up on their prey, others use sheer speed, while still others wait for their food to come to them. To protect themselves from being eaten, some reptiles hide, while others put on a display.

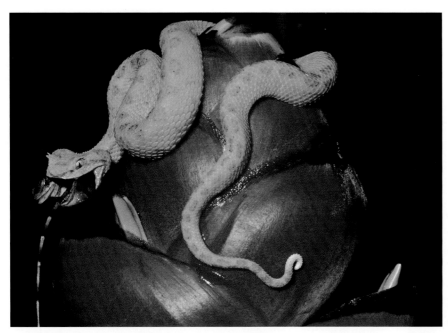

After lying in wait atop a heliconia plant, a tropical eyelash viper catches a small anole lizard and won't need to eat again for several days.

A king cobra swallows a rat snake. Cobras rear up and strike with their fangs, injecting venom into their prey.

If attacked, a Mexican hognose snake rolls over and plays dead to discourage predators.

Alligators often hunt by sneaking up on their prey. They stay low in the water with only snout and eyes exposed.

Taking cover in its hinged shell, a box turtle protects itself from a hungry weasel.

A quick pounce lands this collared lizard a darkling beetle for dinner.

Poised to strike, a prairie rattlesnake loudly warns predators by shaking the tip of its tail. A predator itself, it helps to control rodent populations.

Seeing a reptile like this Nile monitor lizard in the wild is a special experience. Unfortunately, many reptiles are endangered because too many are killed for their meat, skins, or shells, or because their habitats are disappearing. The more we find out about these remarkable animals, the better we can learn how to share the Earth with them.